SDR FOR BEGINNERS

Using the SDRplay and SDRuno

BRIAN SCHELL

SDR FOR BEGINNERS
Using the SDRplay and SDRuno

Copyright 2017-2018 by Brian Schell. All rights reserved, including the right to reproduce this book or any portion of it in any form.

SDRplay and SDRuno are either registered trademarks or trademarks of SDRplay Limited in the United States and/or other countries.

Written and designed by: Brian Schell
brian@brianschell.com

Version Date: July 18, 2018.

ISBN: 1977525806
ISBN-13: 978-1977525802

Printed in the USA of America

Note: The author is not affiliated with SDRplay Limited in any way. He's a ham and a fan of the product. It's not the easiest product to set up, hence this completely unofficial guide.

CONTENTS

Introduction	v
Installing SDRuno	1
Running SDRuno	11
SDRplay and Digital Modes	25
Things to Try	31
About the Author	37
Stay Up To Date!	39
Help Me!	41
Also by Brian Schell	43

INTRODUCTION

SDR is a simple acronym for "Software-Defined Radio." An SDR is a device that includes a full radio tuner in a "black box" with few or no external controls. All the tuning and output must be controlled through an external computer. Some of these SDR devices include both a receiver and transmitter, and others are simpler, containing only the receiver. This book covers the installation, setup, and operation of one particular SDR device, the SDRplay, and the manufacturer's version of the SDR software, called SDRuno.

The SDRplay is a receiver only. It has a huge range of frequencies available, including those for amateur radio, broadcast radio, satellite communication, TV, microwave, and a vast array of other devices. Being receive-only, no special licenses are required; anyone with appropriate computer equipment can buy and use one of these.

I have chosen to describe the operation of the SDRuno software here, since that is the software owned and maintained by the same company that makes the SDRplay, and therefore is sort of the "default" software package. Keep in mind that there are many other SDR control systems out there. Some are quite expensive, and others are free and open-source. I would recommend downloading and installing SDRuno to follow along with the examples here, but once you have everything working and know how things are supposed to behave, it's fun to experiment with other systems.

Why SDR?

Radio operators, both amateur and professional, have been using radios of one form or another for more than a hundred years, and until recent years, they all worked more or less the same way, with physical controls. You turned an analog knob or pressed a digital button or something along those lines to tune the radio to a certain frequency. Whether the tuning controls displayed on an analog meter or a digital readout, they all had one limitation: the operator could only look at one frequency at a time. Later on, dual-frequency VFOs came along, but even then, there were only two frequencies available at once.

There are numerous benefits to a software-defined radio, but one of the main pluses is that the operator can view a range of frequencies, or even an entire band, on the screen at one time. Rather than continuously turning the knob to find a station, or waiting for a digital scanning algorithm to find something to listen to, a listener can watch the spectrum graph and see exactly where the action is across a range of frequencies, clicking on a visual representation of the signal to hear what is being said.

Another benefit of SDR is that all the interface elements are in software, and therefore easily upgradeable as new features and options are added or bugs are removed. If you get tired of the interface and options offered by one control interface software, you can switch to another. Through some open-source options, such as GNU-RADIO, it's possible to program/code your own customized SDR applications. That's not for everyone, but it is is possible to do some really fascinating custom projects this way.

SDR, or more specifically the SDRplay device we'll be covering here also allows easy exploration of voice and non-voice radio applications. The SDRplay RSP1 covers the radio spectrum from 10kHz (Long Wave) to 2GHz (Microwaves), which is a HUGE frequency range. While searching the spectrum with the SDRplay, one can find satellite frequencies, broadcast radio, Citizen's Band, shortwave, some forms of TV, remote control frequencies, commercial handheld radios, all the various voice, packet, and digital modes that are available to the HAM today, as well as modes that often require special hardware, such as D-Star and DMR.

If it's sent by radio, you will have access to it. Decoding the signals is sometimes a challenge, but FINDING the signal isn't.

Cost

Software-defined radios are a fairly-recent development, and as such are still somewhat expensive. It's not hard to find a good SDR transceiver that runs upwards of $3,000.00 USD. The primary benefit of the SDRplay in particular is that it's extremely inexpensive, and it's a good place to try out SDR to see if it's for you. At less than $100.00 in the USA, it's an excellent choice as a "first SDR system" to play with and experiment on. The SDRplay team offers excellent ongoing support, and the receive quality of the SDRplay device itself is outstanding, competing favorably with equipment costing ten times as much. The main "drawback" of the SDR is that it's receive-only, but it has so many other capabilities that it's still a very useful tool to have.

What Do I Need?

The SDRplay alone is just a black box, and doesn't do anything "by itself." It's meant to plug into a computer, although it's very flexible about just what kind of "computer" you need. A PC, Mac, or Linux system will do nicely, but it's even possible to use on a Raspberry Pi or certain tablets and phones. As with any computer application, the more powerful the processor, the more options and capabilities you will have, so a relatively high-power system is recommended.

This is radio after all, so in addition to the computer, you will still need an antenna, and the type of antenna will determine to some extent what you can hear. Just as with a regular radio, you can use anything from a 160-meter long wire to a small handheld-radio-sized "rubber ducky" antenna. Since it's a receive-only device, it's also perfectly fine to experiment with bed springs, a Slinky, or an old satellite dish. The SDRplay comes with a small SMA antenna plug, so you may need an adapter if your antenna uses a larger SO-239 plug. Since it runs on the computer, you will also need to install and configure software, although there are multiple varieties and options that will work.

Although these devices are most commonly used by amateur radio operator enthusiasts (HAMs), a license is not required to use an SDRplay. An FCC license is only required for transmitting on a radio, and since the SDRplay doesn't transmit, it's legal for anyone to purchase and operate. Keep in mind when dealing with antennas, that you need to be aware of general electrical safety issues. Be very aware of overhead power lines. Obviously, lightning and other electrical sources can be dangerous, so be careful where you put that antenna!

Which SDRplay Do I need?

At this writing, there are two models of SDRplay, the RSP1 ($99 USD) and the RSP2($169 USD). They are essentially the same, but the RSP1 has a single antenna port, while the RSP2 has ports for up to three antennas. If you think you may benefit from the ability to have multiple antennas plugged in at once (to tune into two frequencies at the same time), then maybe you want the RSP2. For most users, the RSP1 and its single plug will be fine. There is also a device called the RSP2 Pro($199), which is the same as the RSP2, but has a little bit higher precision for scientific applications. I will be using the RSP1 for all examples here, but the instructions should be the same for the higher-end models.

INSTALLING SDRUNO

I'll be using screenshots and examples from my PC Laptop running Windows 10. Except for the parts where the hardware device is detected and drivers are installed, older versions of Windows should be mostly the same. With Mac software, the specifics will be different, but the general process is going to be similar.

Start up the computer and any web browser, then navigate to SDRplay.com:

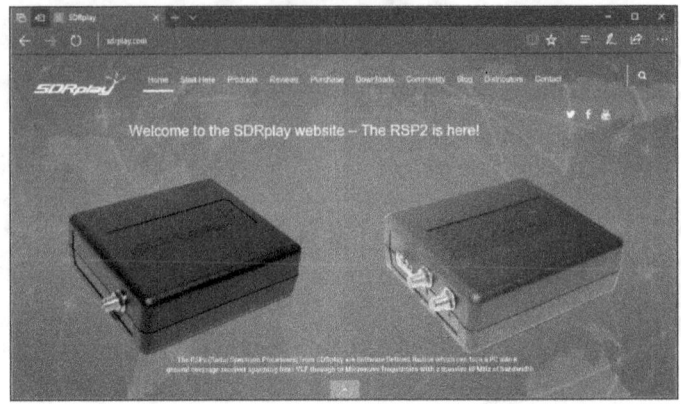

SDRplay Website

Click on the link marked "Start Here."

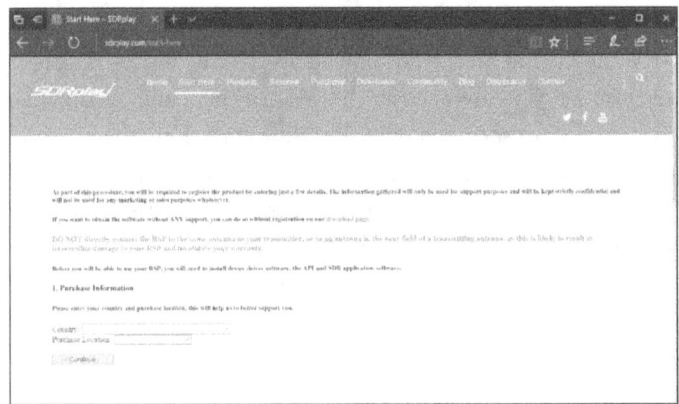

Start Here

Then select your operating system. As you can see below, there are options for Windows, Mac, Linux, Raspberry Pi, and Android, and each one has choices depending on the version of the operating system you use:

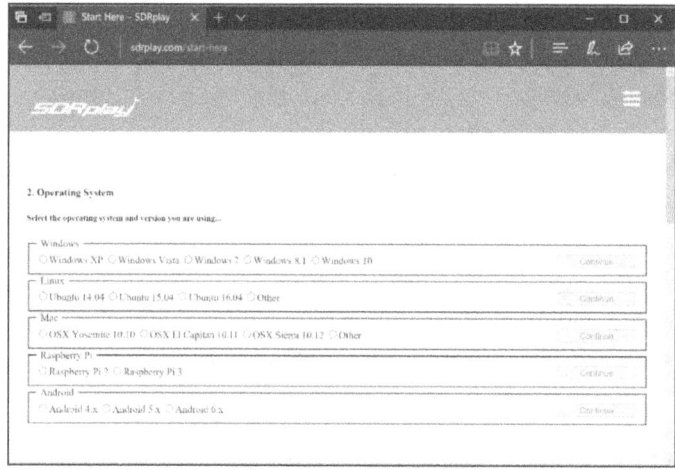

Operating System

Next comes a list of last-minute things to check on. The most important thing at this point is to make sure the RSP cable is NOT connected. If it is already connected, unplug it, as that will mess things up pretty quickly:

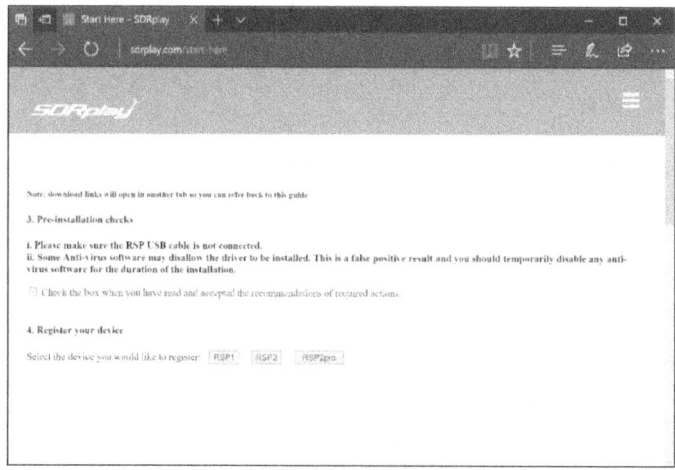

Installation Type

It will also ask for your product serial number. I haven't included a picture of mine, but it's on a sticker underneath the RSP device.

Now you need to decide which software you want to download. The screenshot below shows the various choices for the Windows operating system. If you are running Mac, Linux, or something else, your screen will differ. Also, be aware that these are only the apps that the company has tested and know works well, but other software can work with SDRplay as well. They just don't offer them for download here.

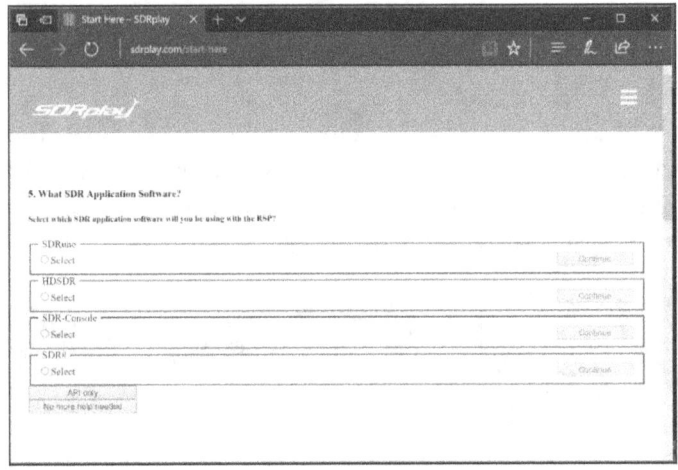

Which Software?

I'm going to choose SDRuno to install and operate throughout the course of this book. It's the "main" software that SDRplay supports, so I'm going to use it here for all examples. The others are all excellent too, but we have to start somewhere. You can always come back here and download the others later.

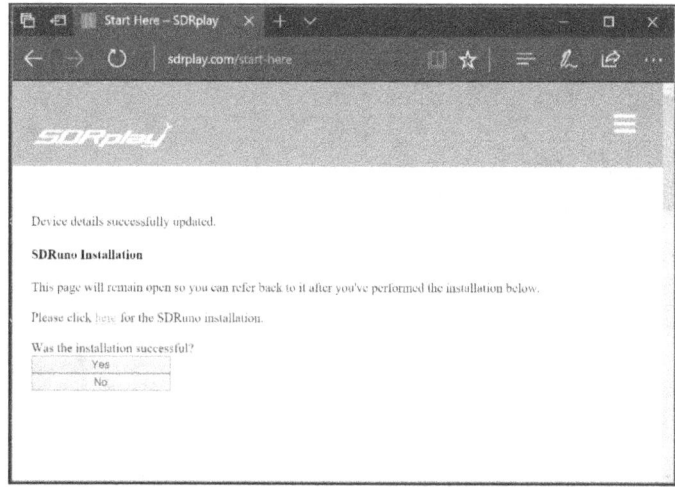

Ready to Download

Click on the link to download and install the software. Again, make sure the SDRplay device is NOT plugged in at this point.

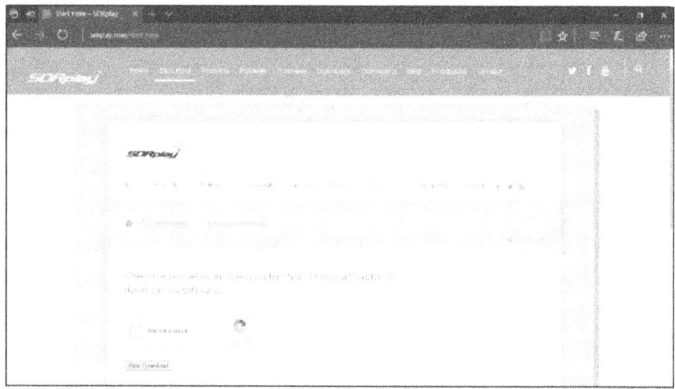

Captcha

Once it's finished downloading, find the downloaded file in the File Explorer and double-click on it. It should pop up a warning and ask if you want to allow it to make changes to your computer. Since the whole point is to install the software, choose "OK" and move on.

Next up is a very standard-looking license agreement. Click on "I AGREE" and NEXT, and once more, you'll get a reminder that your device needs to be UNPLUGGED from your computer:

Warning

Once you click next, it'll ask where you want to install the software. Unless you have some special reason, I would just go with the default:

C:\Program Files (x86)\SDRplay\SDRuno

You will also get a screen asking if you want Start Menu shortcuts and Desktop shortcuts for the software. You will then get a final confirmation screen:

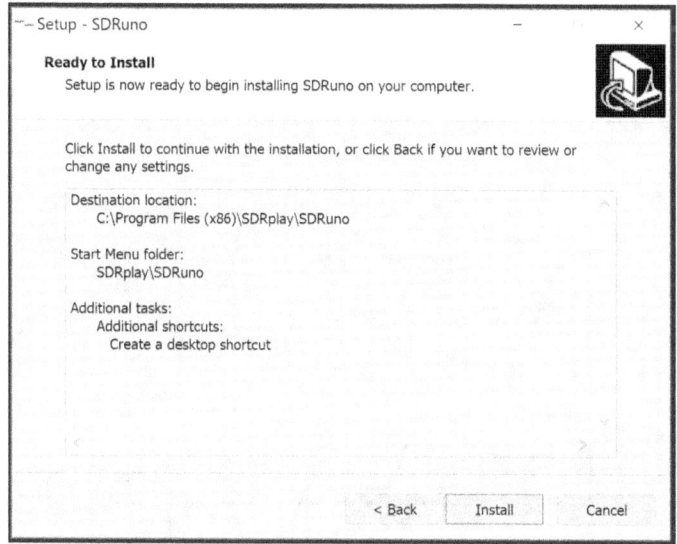

Installation

Click on "Install" and everything should begin. The installation progress bar will fill up and at some point, you will get a notice to plug in your SDRplay device. Plug it in and wait a minute or two. If everything works as it should, the SDRplay should be recognized by Windows, and an appropriate driver will install. If you look at your "Devices" screen, you should see something like this:

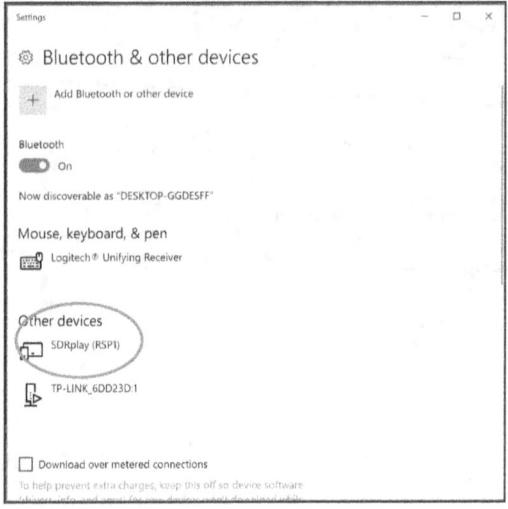

Hardware Devices

Your screen will undoubtedly be different, but as you can see above, the SDRplay (RSP1) shows up as a device.

Now it's OK to click "Next" on the setup program:

Connect

Setup Complete

Now, hold your breath and click "Finish."

If all went well, the installer will go away, and SDRuno should start, leaving you with a window that looks like this:

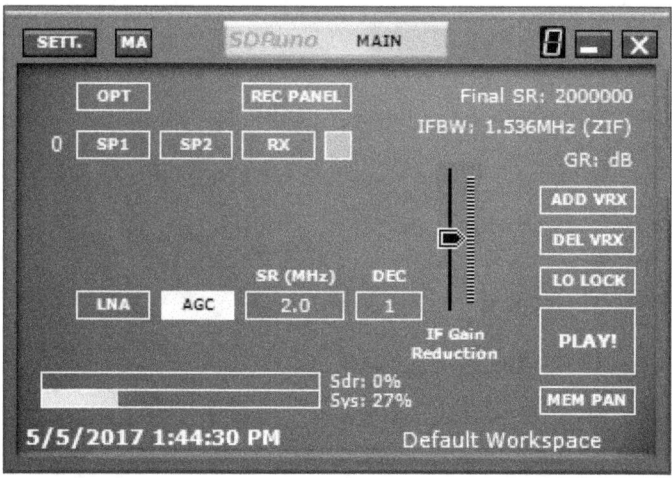

Main SDRplay Window

Let that held breath out. The installation is completed, and it's time to move on to the operation side of things!

RUNNING SDRUNO

Here's a typical screen layout for the main windows that I use with SDRuno. I use the "Main" window (top left), the "RX Control" (top right) and the "Main SP" window (the big one at the bottom). We'll look at each of these individually in this section.

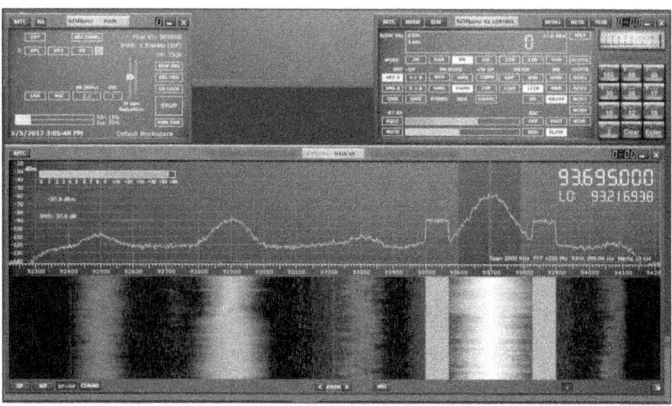

Typical Whole-Screen Layout

Here is the "Main" Window:

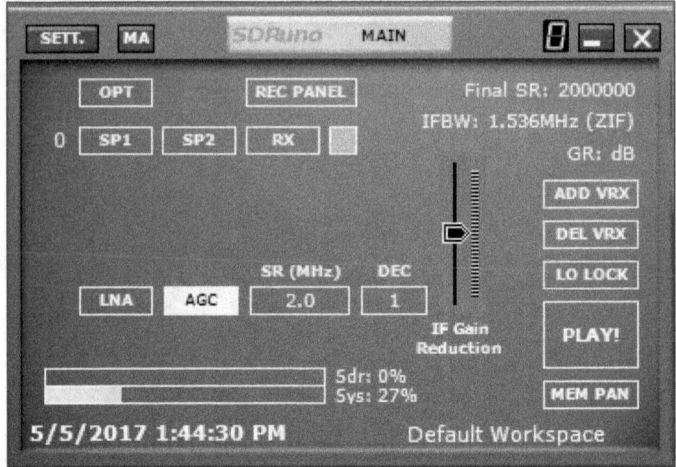

Main Window

In the upper-left of the Main window is a button marked "SETT." This is where many of the more technical hardware settings can be modified or altered. Most of these are pretty technical, and none of them need to be changed to make the SDRplay work, so I will leave them to you to change at your own risk.

Main Settings Options – Six Different Screens

SDR FOR BEGINNERS

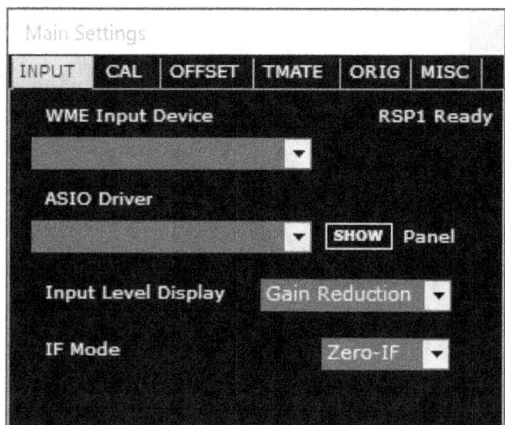

Input Tab

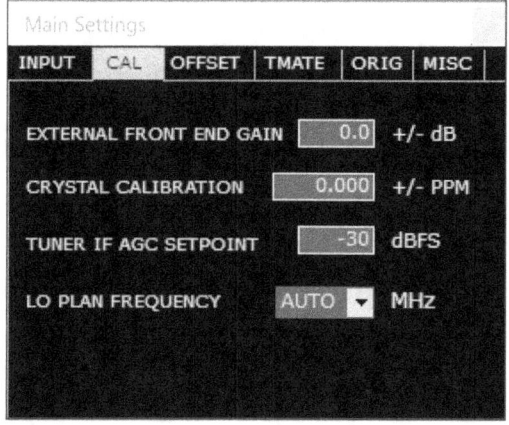

Cal Tab

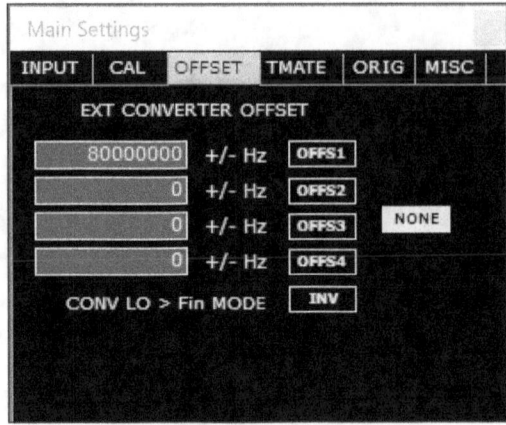

Offset Tab

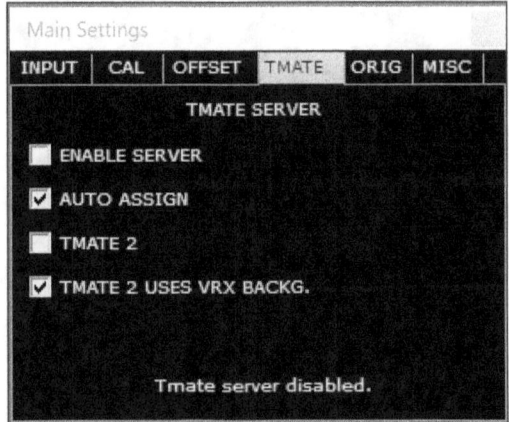

TMATE Tab

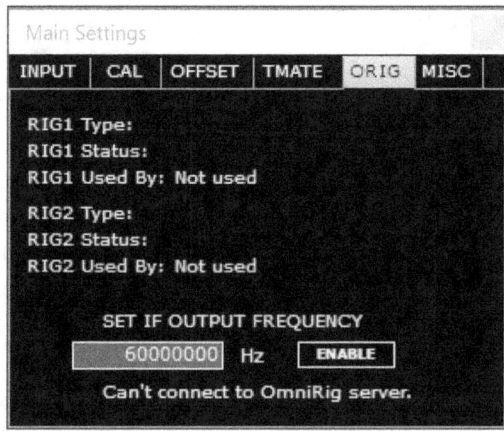

Orig Tab

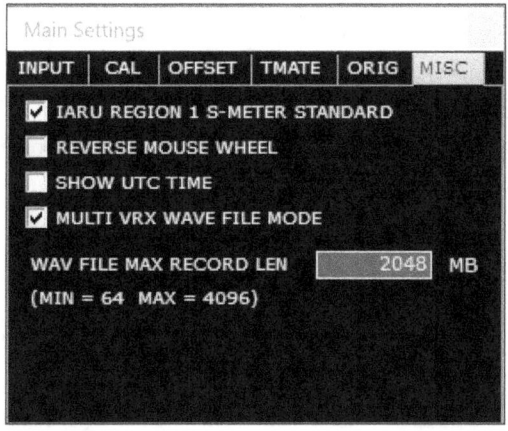

Misc Tab

Back to the Main window, the button marked "OPT" includes many options that are worth experimenting with:

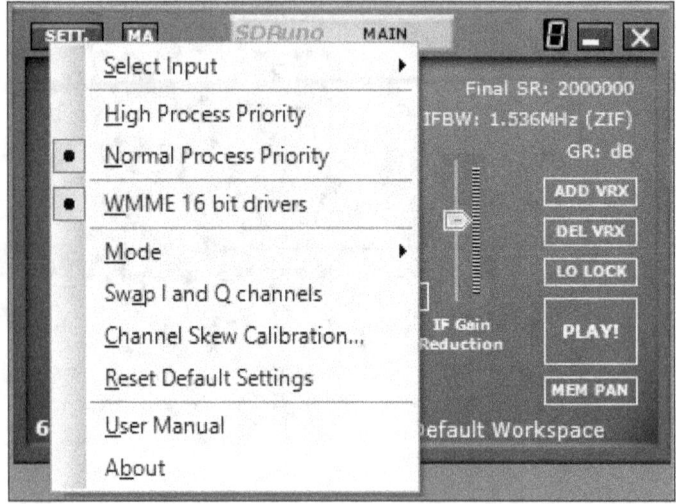

Select Input

Select Input gives you the choice of using either the SDRplay or a WAV file as input. Obviously, most of the time you would want to use the SDRplay. It is useful sometimes to record audio and look at it later, and this is how you would do that. The two **Priority** options can be changed if your machine is spending too much time running some other background process and you want it to focus on SDRuno using more processing power.

The **Mode** option allows you to turn off either the left or right audio channel (the default is that they are both ON). The User Manual is just that: a hyperlink to the SDRuno instructions on the web. The book you hold in your hands is a "For Beginners" guide, and it will get you up and running with all the information you need to know to get started, but there are vast number of options available to you, and the online User Manual is your key to further exploration.

The REC button brings up the Recording window:

Recording Controls

Using this one should be easy to work out, as it uses standard recording controls. (Okay, just in case you need it, top row from left to right is Play, Pause, Repeat. Bottom row from left to right is Stop, Rewind, Record).

Clicking on the SP1 button from the Main windows brings up the Spectrum window:

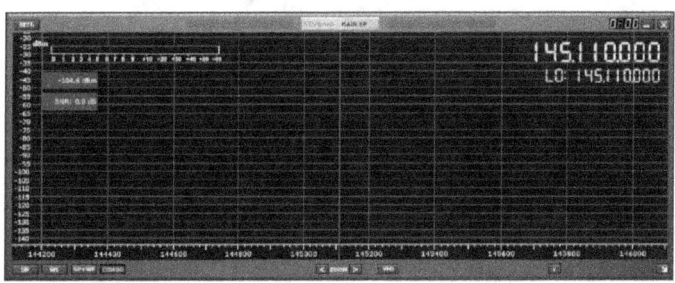

Spectrum Controls

At the bottom of the spectrum windows are four buttons: SP, WF, SP+WF, and Combo. These give you various combinations of the Spectrum Window and the Wave Form window. Try all four and see for yourself what they do. The zoom arrows in the bottom center zoom in or out on the spectrum, allowing you to fine-tune the radio more carefully. At the top-left of the Spectrum window is another "SETT." button for settings:

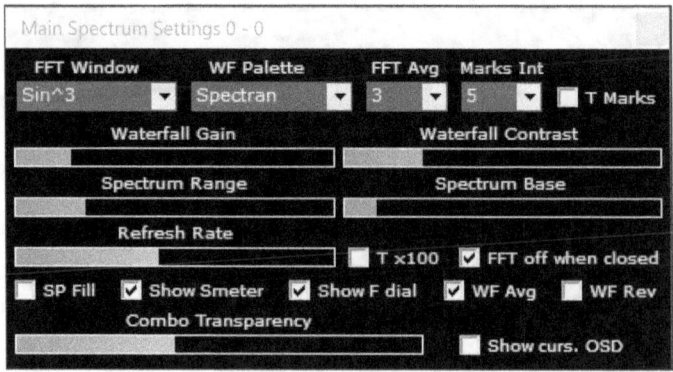

Spectrum Settings

For the most part, these settings pertain to the appearance of the waterfall and spectrum, and you can experiment with these until everything looks good for you.

The next important button on the Main window is the one marked "Play!" it's time to click this one. This is the equivalent of turning on the power with a regular radio. Assuming the hardware is connected and the drivers are working, you should see some activity on the Waveform screen and hear something (probably just static until you tune it) from your audio device/speakers:

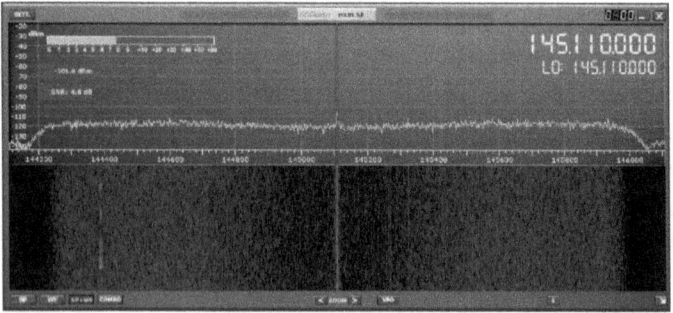

Waveform

Going back to the Main window, the "RX" button brings up the most complex-looking window so far, the receiver, or "RX Control."

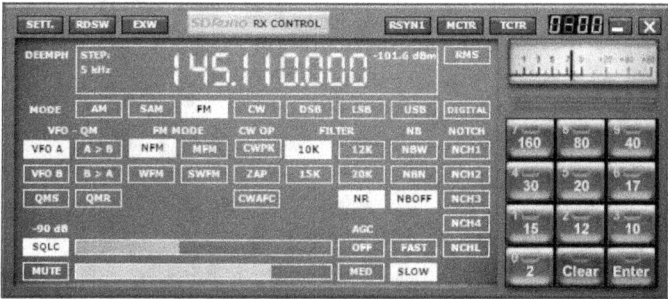

RX Control

This is where most of your choices will come into play. Many of the controls and options here will look similar to what you would find on a physical radio, with buttons for VFO, AM, FM, USB, Squelch, AGC, Noise Blockers, and other functions that are common on modern amateur radios.

On the top bar of the RX Control window is a button for "EXW". This brings up the "Extra Controls" Window:

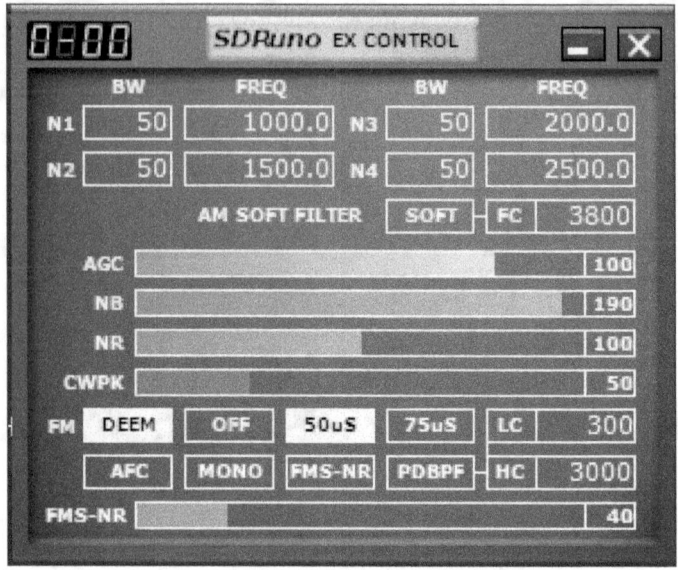

EX Control

There are some advanced settings here that we will get into later, but for now, you can skip these.

You can resize the windows and drag them around for easier access. I prefer the Main, RX, EX, and Spectrum windows open at the same time, like this:

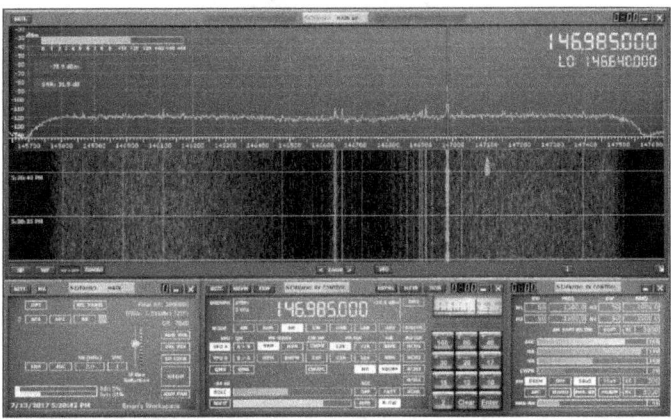

All Windows, Arranged

It quickly gets to be a pain opening and moving all these windows around every time you start the program, so it's good to save the layout. This can be done from the Main window, down at the bottom-right where it says, "Default Workspace." In the picture above, mine says "Brian's Workspace." Left-click in it, then type in whatever you want to name it. Then hold down the CTRL key and click it again, this time choosing which "slot" you want to save in. It's a little non-intuitive, but once you've got it, it will save you time later.

Once you have the windows open and arranged how you like, it's time to turn it all on. Click the "Play!" button in the Main window if you haven't already. You'll probably hear static, so adjust the volume in your speakers or headset so it's not too loud or uncomfortable.

You can hook the SDRplay up to most any kind of antenna and receive signals from the appropriate frequencies by choosing an amateur radio band by clicking on the "keypad" on the right side of the RX Control window.

For now, let's focus on broadcast FM radio stations for a moment, because they're nearly everywhere and easy to find strong signals. In the RX Control window, the frequency appears near the top.

You can change the frequency by pointing at the individual digits and using the mouse wheel to move the values of individual digits up and down. It seems slow until you get used to it, but the more you do it, the easier it gets. In my example screenshots, I have tuned to a local music station, 107.7 FM:

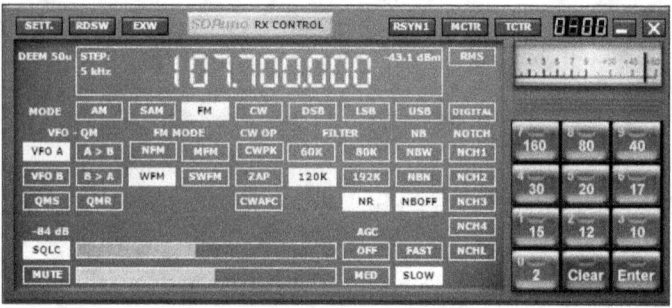

RX Control

You'll see the signal in the SP window change as well as the waterfall. Assuming you live anywhere near a city, the FM broadcast band is usually busy and colorful:

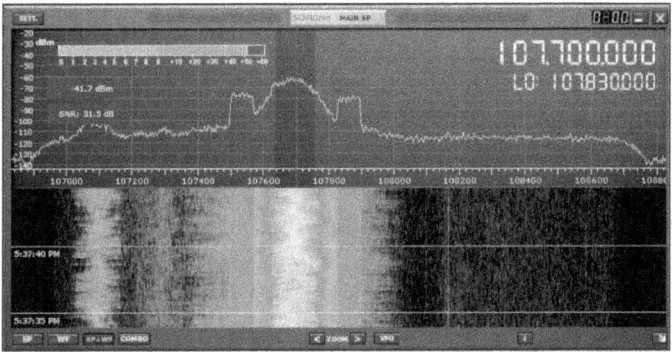

FM Radio

In the RX Control window, there are several options for "FM Mode" including NFM (Narrow FM), WFM (Wide FM), MFM (Medium FM), and SWFM Stereo Wide FM. Each of these sets the bandwidth to a preset amount. For stereo FM, the bandwidth is wider than most modes, so clicking the SWFM button sets the bandwidth to the widest. You can visually see how wide it is on the Main SP window. Click on each of the FM settings buttons to see what happens. For something narrower, like SSB voice, a tighter setting would be more appropriate. Listening to a Morse code signal would benefit from the narrowest bandwidth setting. Once you have it set on SWM, click on the "192K" button for selectivity filter. Assuming the signal is strong, any noise should clear right up.

An interesting feature of SDRuno that comes into play with broadcast stations is that it can decode RDS (Radio Data System). This is digital data encoded within the audio signal that allows some radio receivers to display what song is playing, who the artist is, etc. If you have a nice, clear FM signal coming through (this probably only works for music on most stations), click the "RDSW" button up at the top of the RX Control window. Assuming the station you have tuned in uses RDS (they don't all support it), a window will pop up showing data for the current song:

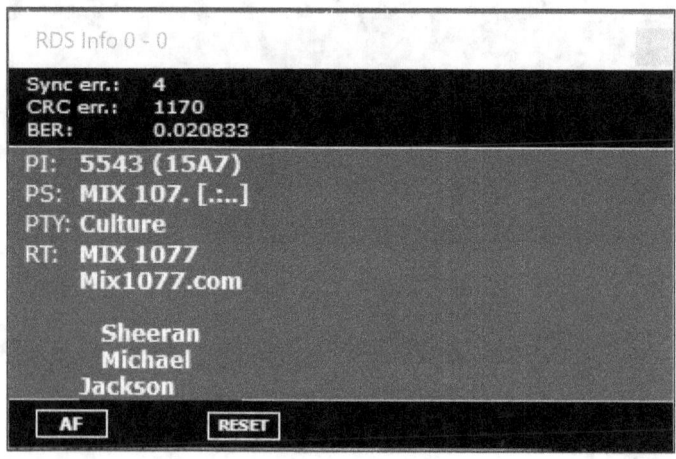

Station Info

Of course, you probably didn't buy the SDRplay to listen to FM broadcast radio, but the process and controls are the same for about any signal. Set the frequency, bandwidth, any kind of filter or mode you need, and then just listen. One of the strengths of SDR is that if you don't *know* what frequency you want, you can look at the graphic representation of the band in the Main SP window and click on the various peaks to see what you get.

Play around with it. Hover your mouse over the various buttons to get an idea what they all do. To learn what all those buttons and settings do, don't forget that SDRuno has an excellent online instruction manual that explains what every button and setting does.

Now on to more advanced usage.

SDRPLAY AND DIGITAL MODES

Listening to voices, music, and audio is relatively easy on SDRplay / SDRuno, as we have just seen. A lot of amateur radio enthusiasts like to use digital modes such as PSK31, APRS, RTTY, JT65/JT9, and uncountable others. Can you listen to those through SDRplay? Absolutely!

First, you will need to download and install whatever decoding software you want that supports the desired digital mode: HRD or Fldigi or WSJT-X or something like that.

With a "regular radio," you would just get the software running on your computer that decodes the mode you are interested in, and then you plug the audio output from your radio into the microphone or line input on your computer. Sometimes some kind of hardware interface is required to make that connection possible. Then you could decode any signal you wanted.

If you've ever done it on a regular transmitter, you know it's never actually that simple, and SDRplay is no different—there's a catch. The problem is that the SDRuno software is designed to play sound on the computer's audio output device, that means headphones or a speaker. The digital decoding software (whatever type you use) tries to get its input from the microphone or line input. The problem lies in that most computers can't accept audio input while they are outputting sound. Just like with most people, when it tries to "talk" and "listen" at the same time, it can't do either well.

The "brute force approach" to solving this problem is to use two computers, one for the decoding software, and one for the SDRplay. In a perfect world, that would work fine. Most of us don't have a second computer or just don't want to devote two computers to this, so that's not the most efficient solution.

Fortunately, there is software out there that can "fool" the computer into doing exactly what we want.

For Windows, **Virtual Audio Cable** is a special software "driver" that you can set up that does exactly what we need for this. You install the VAC driver, then set the SDRuno software to output to that "virtual device" that tricks Windows into thinking it is a sound output device. Then you tell your digital decoding software to get its input from that same virtual device. There are no *real* patch cables or devices needed for this, and if this is all you wanted the RSPlay for, you could in fact do this without even having a sound card in your PC.

If you are running a Mac, there is a similar app available called **Audio Hijack Pro** that is very similar to VAC.

The flow goes like this:

SDRplay Device -> SDRuno Software -> Virtual Audio Cable -> WSJT-X (or some other decoding software)

Here is an image of the "Audio Settings" window for

WSJT-X, a common application that decodes JT9/JT65 audio into readable messages. Note that the audio INPUT comes from the VAC driver:

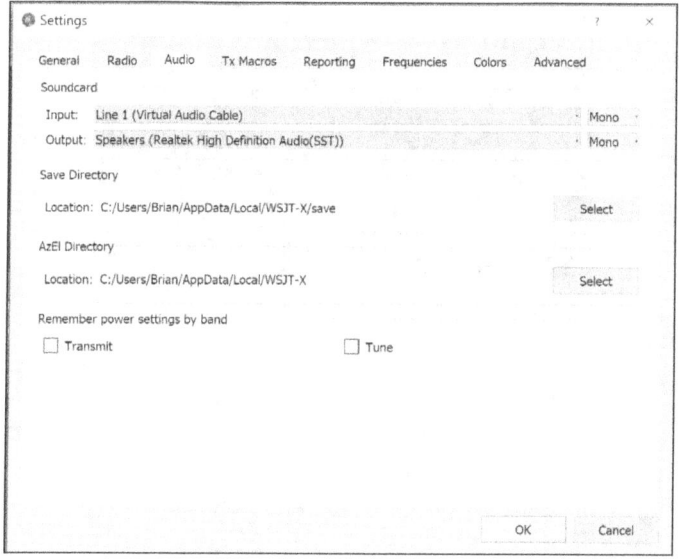

Virtual Audio Cable Settings

And to match up with that, you set the Audio Output of SDRuno to use the VAC as well. This is done in the "Settings" menu of the "RX Control" window:

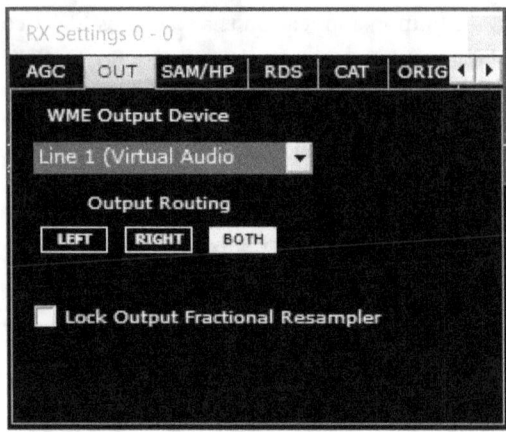

RX Control Settings

Once you have the SDRuno output going to the Virtual Audio Cable, and the input of the WSJT-X (or whatever software you need) set to listen to the VAC, then just follow the instructions for the decoding software, and the rest should be straightforward!

Virtual Audio Cable for Windows is available at http://software.muzychenko.net/eng/vac.htm

Audio Hijack Pro for Mac can be found at https://www.rogueamoeba.com/audiohijackpro/

Neither of these are free, but they each have a free trial that you can try before purchasing.

Notch Filter

It's not unusual to tune into a signal that seems strong, but it is hard to hear because something else is transmitting on a nearby frequency, drowning it out. A "notch filter" can solve this problem.

1. Tune into the frequency you want and try to determine the section of the waveform where the "noise" is coming from.
2. Click "SP2" from the Main Window to bring up the Aux SP window. This window shows a zoomed-in area of the band.
3. Click "NCH1" on the RX Control window. You won't see anything happen yet.
4. SHIFT-Left Click in the Aux SP window at the point where the "noise" or offending signal is located.
5. In the SP window, you can now see that spot on the waveform is "zeroed out" and there's a "black stripe" in the waterfall beneath it, like this:

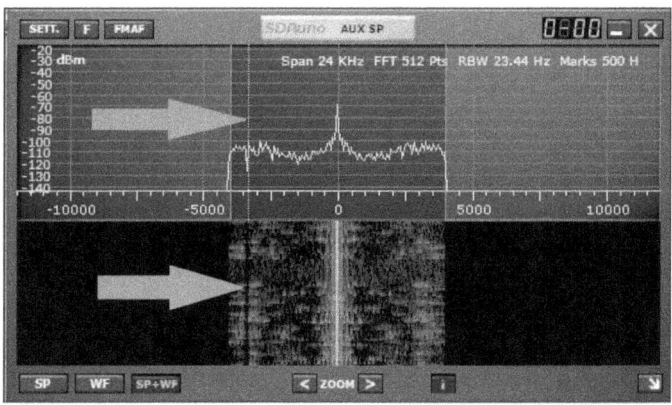

Notch Filter

And assuming everything went well, the offending sound has been blanked out. Occasionally, you may need to place TWO notches for a signal where there are two extraneous noises. Do the same process again, only this time click the NCH2 button on the main screen and use Shift *right*-click to place the notch in the Aux SP window.

Notch Filter Placement:
Shift Left-Click for NCH1
Shift Right-Click for NCH2

THINGS TO TRY

Listen to D-Star and/or DMR radio:

1. This requires both the Virtual Audio Cable software described earlier and the DSD Decoder found at http://dsdplus.com
2. Set the audio output from SDRuno to send its output to VAC. This is done the same way described in the section concerning digital modes. The DSDPlus software should accept input from VAC automatically.
3. Find out what the frequency is for your local D-Star or DMR repeater (The process is the same for either).
4. Tune to that frequency.
5. Change your settings in SDRuno to FM mode, 15K Filter. Also, in the EX Control window, make sure "DEEM" is OFF:
6. Start the DSDPlus app. Two status windows will open. You can hit the ? key on your keyboard to

see what commands and switches you can set while these run.
7. Wait on someone to make a call.
8. Remember that the DSDPlus software is someone's hobby project, and although it works on many repeaters and systems, your mileage may vary.

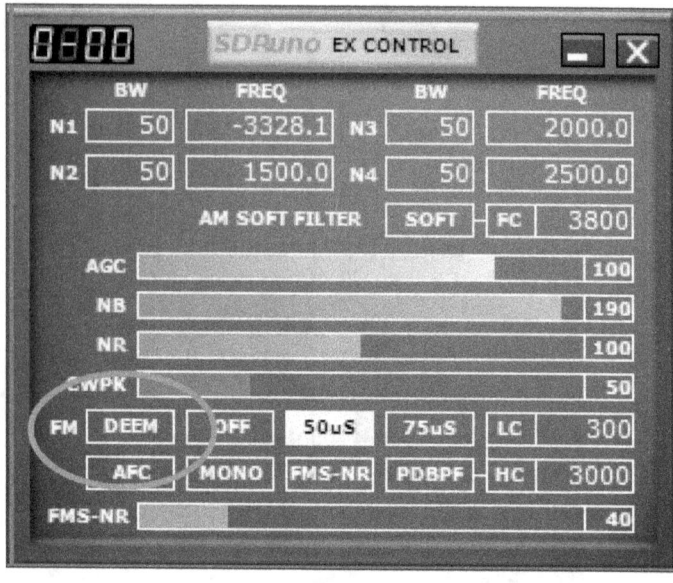

DEEM (De-Emphasize) is Off

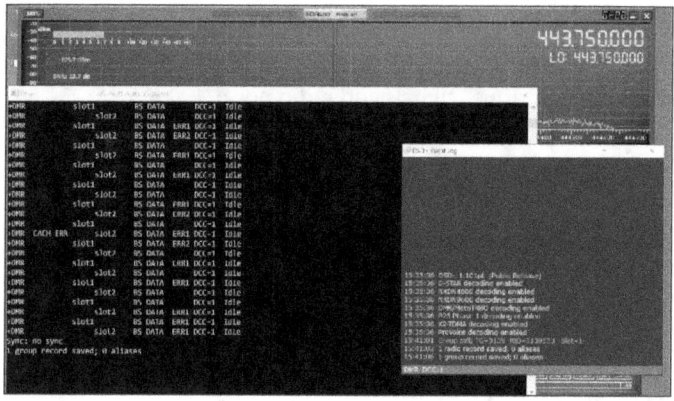

DSDPlus for D-Star and DM

Additional Links for Advanced Topics

Watch TV

It's a real challenge, and it's pretty limited, but it's possible. Here's a link to a user who made it work:

http://coolsdrstuff.blogspot.-
com/2015/09/watching-atsc-hdtv-on-sdrplay-
rsp.html

Save the waterfall as a time lapse:

This one's also a bit of a challenge, but it's also possible to do it. If you ever wanted to know what the busiest frequencies are in your area, this will show you visually where the action is. It can also alert you to appliances and electronics that are emitting "spurious" signals and radio noise.

```
http://coolsdrstuff.blogspot.-
com/2016/08/long-term-waterfall-on-
sdrplay.html
```

Try using a satellite dish:

This one's easy if you have access to a satellite dish. Just plug it in and see what you can find on the waterfall. There's a reason HAM operators generally don't use dishes, but that doesn't mean you can't experiment with them.

```
http://coolsdrstuff.blogspot.-
com/2016/12/5ft-15m-satellite-dish.html
```

The Official Website

And of course, and most importantly, check out the SDRplay website. They post video tutorials and helpful tips on their blog regularly. Several of the tips in this book came from watching their videos, and there's a lot to learn there.

```
http://www.sdrplay.com/blog/
```

http://www.youtube.com/channel/UC4JDq3US2eb1N4dRCT45_Zw

ABOUT THE AUTHOR

Brian Schell (KD8OTD) is a former College IT Instructor who has an extensive background in computers dating back to the 1980s. Currently, he writes on a wide array of topics from computers, to world religions, to ham radio, and even releases the occasional short horror tale.

He'd love to hear your stories of success and failure with SDRplay. If there's something you would like to see in a future edition of the book, or otherwise have suggestions, please drop him a note. Contact him at:

```
Web: http://BrianSchell.com
Email: brian@brianschell.com
```

- twitter.com/BrianSchell
- facebook.com/Brian.Schell
- instagram.com/brian_schell
- pinterest.com/brianschell

STAY UP TO DATE!

Join my email update list— There's NO weekly SPAM or filler material, only announcements of new books or major updates.

http://brianschell.com/list/

HELP ME!

CONTACT THE AUTHOR

If you have a suggestion or find a mistake, email me about it, and I'll get it into the next edition of the book. Got a gripe, complaint, question, or just adoring fan mail? Same thing!

LEAVE A REVIEW

If this book helped you, please leave a review where you purchased this book. Reviews are the best way to help out!

SHARE WITH YOUR FRIENDS

Did you enjoy this book? Please use the buttons below to spread the word to your friends and followers.

ALSO BY BRIAN SCHELL

Amateur Radio

- D-Star for Beginners
- Echolink for Beginners
- DMR for Beginners Using the Tytera MD-380
- SDR for Beginners with the SDRPlay
- OpenSPOT for Beginners
- Programming Amateur Radios with CHIRP
- FM Satellite Communications for Beginners

Technology

- Going Chromebook: Living in the Cloud
- Going Text: Mastering the Power of the Command Line
- Going iPad: Ditching the Desktop
- DOS Today: Running Vintage MS-DOS Games and Apps on a Modern Computer

Old-Time Radio Listener's Guides

- OTR Listener's Guide to Dark Fantasy

The Five-Minute Buddhist Series

- The Five-Minute Buddhist
- The Five-Minute Buddhist Returns
- The Five-Minute Buddhist Meditates
- The Five-Minute Buddhist's Quick Start Guide to Buddhism

- Teaching and Learning in Japan: An English Teacher Abroad

Fiction with Kevin L. Knights:
- Tales to Make You Shiver
- Tales to Make You Shiver 2
- Random Acts of Cloning
- Jess and the Monsters

www.ingramcontent.com/pod-product-compliance
Lightning Source LLC
Chambersburg PA
CBHW050024230526
45470CB00003B/1114